FREQUENCY WAVE THEORY

Expands on ERIC WEINSTEIN Theories

By Drew Ponder

PREFACE

The quest to understand the fundamental nature of reality has captivated humanity for millennia. From the ancient philosophers who pondered the mysteries of existence to the modern scientists who probe the depths of the cosmos, our journey has been one of discovery, curiosity, and profound insight. This book explores a groundbreaking integration of two powerful theoretical frameworks: Eric Weinstein's Geometric Unity and Drew Ponder's Frequency Wave Theory. Through their synthesis, we aim to unlock new perspectives on the universe, its underlying principles, and our place within it.

Geometric Unity seeks to bridge the seemingly insurmountable gap between general relativity and quantum mechanics, proposing additional dimensions and intricate geometric structures. Frequency Wave Theory, on the other hand, posits that all matter and energy are manifestations of underlying vibrational patterns, governed by principles of frequency and resonance. Together, these theories offer a unified model that transcends traditional boundaries and promises to revolutionize our understanding of the cosmos.

As we delve into this integrated framework, we will explore its mathematical foundations, experimental evidence, and wide-ranging applications. From the development of innovative technologies and advancements in healthcare to the promotion of sustainability and social equity, the implications of this unified model are vast and far-reaching. We will also consider the philosophical and metaphysical questions that arise from this synthesis, challenging us to rethink our concepts of existence, consciousness, and interconnectedness.

This book is intended for a diverse audience: scientists and engineers seeking new theoretical and practical insights, educators and students eager to explore cutting-edge ideas, and anyone with a deep interest in the mysteries of the universe. By presenting these complex concepts in an accessible and interdisciplinary manner, we hope to inspire curiosity, foster collaboration, and ignite a passion for discovery.

As we embark on this journey together, let us keep an open mind and a spirit of inquiry. The integration of Geometric Unity and Frequency Wave Theory offers a powerful lens through which to view the world, one that emphasizes the harmony and interconnectedness of all things. Through this lens, we can uncover new truths, develop transformative technologies, and ultimately, contribute to a more enlightened and harmonious future.

Welcome to the exploration of a unified theory that has the potential to reshape our understanding of the universe and our place within it. May this journey inspire you, challenge you, and open your mind to the infinite possibilities that lie ahead.

CHAPTER 1

The Convergence of Theories

I n the quest for understanding the fundamental nature of
reality, the convergence of innovative theories can often
lead to groundbreaking discoveries. This book explores the
synthesis of Eric Weinstein's Geometric Unity with the Frequency
Wave Theory, creating a novel framework that seeks to unify
the disparate realms of physics, cosmology, and beyond. By
integrating these powerful ideas, we embark on a journey to
uncover new insights into the universe's deepest mysteries and
their implications for our world.

The Foundation of Geometric Unity

Eric Weinstein's Geometric Unity represents a bold attempt to address the long-standing challenges in theoretical physics. At its core, this framework aims to reconcile general relativity and quantum mechanics, two pillars of modern physics that have remained stubbornly incompatible. Geometric Unity proposes additional dimensions and intricate mathematical structures, suggesting that the universe's fabric is far more complex than previously imagined.

Weinstein's approach leverages advanced differential geometry and gauge theory, envisioning a unified field that could elegantly encompass gravity, electromagnetism, and the strong and weak nuclear forces. This ambitious theory challenges the dominance of string theory, offering a fresh perspective that could potentially unlock new avenues of understanding.

The Essence of Frequency Wave Theory

Frequency Wave Theory, conceptualized by Drew Ponder, introduces a complementary paradigm focused on the principles of frequency, vibration, and resonance. This theory posits that all matter and energy in the universe are manifestations of underlying vibrational patterns. By understanding these patterns, we can gain insights into the behavior of particles, forces, and even consciousness.

The applications of Frequency Wave Theory are vast, ranging from medical advancements in targeting cancer cells to innovations in renewable energy through geothermal drilling. This theory also extends into the realms of communication technology, cosmology, and philosophical inquiries into the nature of existence.

The Power of Integration

The integration of Geometric Unity with Frequency Wave Theory holds the promise of a more comprehensive understanding of the universe. While Geometric Unity provides a robust mathematical framework, Frequency Wave Theory introduces a dynamic, vibrational perspective that can enrich our comprehension of physical phenomena. Together, these theories can address unanswered questions in both classical and quantum physics, potentially leading to a unified model that encompasses all fundamental forces.

This chapter sets the stage for our exploration by outlining the foundational concepts of both Geometric Unity and Frequency Wave Theory. As we delve deeper into their integration, we will uncover how the convergence of these ideas can offer new insights into the nature of reality, from the behavior of subatomic particles to the structure of the cosmos.

The Journey Ahead

In the chapters that follow, we will embark on a journey through the intricacies of these theories, examining their implications and applications. We will explore how the principles of frequency and resonance can be woven into the geometric fabric of the universe, creating a cohesive model that transcends traditional boundaries. This journey will not only expand our scientific horizons but also challenge our philosophical and metaphysical understanding of the world.

By the end of this book, we aim to present a unified theory that harmonizes the mathematical elegance of Geometric Unity with the dynamic essence of Frequency Wave Theory. This synthesis will pave the way for future research, technological innovation, and a deeper appreciation of the universe's profound complexity.

Join us as we venture into the convergence of theories, seeking to illuminate the path to a more comprehensive understanding of reality.

CHAPTER 2

Understanding Geometric Unity

I n order to appreciate the full potential of integrating Geometric Unity with Frequency Wave Theory, it is essential to first delve deeply into the foundational principles and ideas behind Eric Weinstein's Geometric Unity. This chapter provides a comprehensive overview of Weinstein's theory, its motivations, and the mathematical constructs that underpin it. By understanding the essence of Geometric Unity, we set the stage for its synthesis with Frequency Wave Theory in subsequent chapters.

The Motivation Behind Geometric Unity

The primary motivation for Geometric Unity arises from the unresolved tensions between general relativity and quantum mechanics. General relativity, formulated by Albert Einstein, describes the macroscopic behavior of gravity and the curvature of spacetime. Quantum mechanics, on the other hand, governs the behavior of particles at the smallest scales. Despite their success in their respective domains, these theories are fundamentally incompatible when it comes to describing phenomena where both gravity and quantum effects are significant, such as in black holes and the early universe.

Weinstein's Geometric Unity aims to bridge this gap by proposing a unified framework that can accommodate both the curvature of spacetime and the probabilistic nature of quantum mechanics. This ambitious goal requires rethinking the very fabric of the universe and introducing new mathematical constructs that can seamlessly integrate these two pillars of modern physics.

The Mathematical Foundations

At the heart of Geometric Unity lies the sophisticated use of differential geometry and gauge theory. Differential geometry provides the language to describe the curvature of spacetime, while gauge theory offers a framework for understanding the fundamental forces of nature through the concept of symmetry.

Additional Dimensions: One of the key features of Geometric Unity is the introduction of additional dimensions beyond the familiar three spatial and one temporal dimension. These extra dimensions are not directly observable but can have profound effects on the behavior of particles and fields in our four-dimensional spacetime. By extending the dimensionality of spacetime, Weinstein's theory can potentially unify the gravitational force with the other fundamental forces.

Fiber Bundles and Gauge Fields: Geometric Unity employs the concept of fiber bundles, which are mathematical structures that allow different types of fields to be consistently defined over a curved spacetime. Gauge fields, which describe the interactions between particles, can be naturally incorporated into this framework. This allows for a more elegant and comprehensive description of the fundamental interactions.

Symmetry and Group Theory: Symmetry plays a central role in Geometric Unity. The theory utilizes advanced group theory to describe how different particles and forces transform under various symmetries. These symmetries are essential for maintaining the consistency and coherence of the unified framework.

Key Predictions and Implications

Geometric Unity is not just a theoretical exercise; it makes several key predictions that can be tested experimentally. Some of these predictions include:

New Particles and Forces: The existence of additional dimensions and new symmetries implies the presence of previously undiscovered particles and forces. These new entities could be detected in high-energy particle accelerators or through astrophysical observations.

Modifications to General Relativity: Geometric Unity suggests modifications to the equations of general relativity at very small scales or in regions of extreme curvature, such as near black holes. These modifications could lead to new insights into the nature of spacetime singularities and the behavior of gravity at quantum scales.

Unification of Forces: Perhaps the most profound implication of Geometric Unity is the potential unification of all fundamental forces into a single coherent framework. This would represent a significant step forward in our understanding of the universe and could open up new avenues for research and technological innovation.

Challenges and Criticisms

Like any groundbreaking theory, Geometric Unity faces several challenges and criticisms. One of the main challenges is the lack of experimental evidence for additional dimensions and the new particles predicted by the theory. Additionally, the mathematical complexity of the theory makes it difficult to derive specific predictions that can be tested with current technology.

Despite these challenges, Geometric Unity represents a bold and innovative approach to one of the most profound questions in physics. By providing a framework that can potentially unify general relativity and quantum mechanics, Weinstein's theory offers a tantalizing glimpse into a deeper understanding of the universe.

Conclusion

Geometric Unity is a visionary theory that seeks to address some of the most fundamental challenges in modern physics. By exploring the intricate mathematical structures and concepts that underpin this theory, we gain a deeper appreciation of its potential to transform our understanding of the universe. In the next chapter, we will explore the principles of Frequency Wave Theory and begin to uncover how these two powerful ideas can be integrated to create a unified model of reality.

CHAPTER 3

*The Fundamentals of
Frequency Wave Theory*

Having explored the foundational concepts of Geometric Unity, we now turn our attention to Frequency Wave Theory. This theory, conceptualized by Drew Ponder, offers a fresh perspective on the nature of reality by emphasizing the roles of frequency, vibration, and resonance. In this chapter, we will delve into the core principles of Frequency Wave Theory, its key applications, and how it complements and enhances Geometric Unity.

The Core Principles of Frequency Wave Theory

At its heart, Frequency Wave Theory posits that all matter and energy in the universe are manifestations of underlying vibrational patterns. These patterns are governed by specific frequencies and resonances, which determine the properties and behaviors of particles, fields, and forces. The theory can be summarized through several key principles:

Vibrational Essence: Everything in the universe, from the smallest subatomic particles to the largest cosmic structures, vibrates at specific frequencies. These vibrations create standing waves that define the fundamental characteristics of matter and energy.

Resonance: Resonance occurs when an object or system is driven by an external force to oscillate at a particular frequency. In Frequency Wave Theory, resonance is a crucial mechanism that explains how different systems interact and transfer energy.

Frequency Domains: The universe can be thought of as a series of interconnected frequency domains, each with its own set of vibrational patterns. These domains interact and overlap, leading to complex behaviors and phenomena observed in nature.

Harmonic Relationships: The frequencies of vibrations are often related by simple harmonic ratios, similar to the harmonics in musical notes. These relationships create a coherent and interconnected structure within the universe.

Key Applications of Frequency Wave Theory

Frequency Wave Theory has a wide range of applications across various fields, offering new insights and solutions to complex problems. Some of the key applications include:

Medical Advancements: One of the most promising applications of Frequency Wave Theory is in the field of medicine. By understanding the specific frequencies at which different cells and tissues vibrate, researchers can develop targeted therapies for diseases such as cancer. For example, resonant frequencies can be used to selectively target and destroy cancer cells without harming healthy tissue.

Renewable Energy: Frequency Wave Theory provides novel approaches to harnessing renewable energy sources. In geothermal energy, for instance, understanding the vibrational properties of the Earth's crust can improve the efficiency of drilling and energy extraction processes.

Communication Technology: The principles of frequency and resonance can enhance modern communication systems, leading to more efficient and reliable transmission of information. This includes advancements in wireless communication, quantum communication, and brain-computer interfaces.

Cosmology and Astrophysics: Frequency Wave Theory offers new perspectives on the behavior of celestial bodies and the structure of the universe. By analyzing the vibrational patterns of stars, galaxies, and other cosmic entities, scientists can gain deeper insights into their formation, evolution, and interactions.

Complementing Geometric Unity

The integration of Frequency Wave Theory with Geometric Unity creates a powerful framework that bridges the gap between macroscopic and microscopic phenomena. While Geometric Unity provides a robust mathematical structure for unifying the fundamental forces, Frequency Wave Theory introduces a dynamic, vibrational perspective that enriches our understanding of physical interactions.

Unified Description of Forces: By incorporating vibrational patterns and resonances into the geometric framework, we can achieve a more holistic description of the fundamental forces. This unified approach can potentially resolve inconsistencies between general relativity and quantum mechanics.

New Predictions and Models: The synthesis of these theories allows for the development of new models and predictions that can be tested experimentally. For instance, the vibrational properties of additional dimensions proposed by Geometric Unity can be explored through the lens of Frequency Wave Theory, leading to potential discoveries of new particles and forces.

Interdisciplinary Insights: The combined framework encourages interdisciplinary research, drawing on insights from physics, mathematics, biology, and engineering. This collaborative approach can foster innovation and lead to breakthroughs in both theoretical and applied sciences.

Conclusion

Frequency Wave Theory offers a transformative perspective on the nature of reality, emphasizing the fundamental roles of frequency, vibration, and resonance. By understanding these principles, we can unlock new applications and insights across various fields. When integrated with Geometric Unity, Frequency Wave Theory enhances our ability to describe and understand the universe in a more comprehensive and coherent manner.

In the next chapter, we will explore the mathematical integration of these theories, examining how the geometric constructs of Geometric Unity can be enriched by the dynamic principles of Frequency Wave Theory. This synthesis will pave the way for a unified model that transcends traditional boundaries and opens up new possibilities for scientific exploration.

CHAPTER 4

*Mathematical Integration
of Geometric Unity and
Frequency Wave Theory*

With a firm understanding of both Geometric Unity and Frequency Wave Theory, we now delve into the intricate process of integrating these two powerful frameworks. This chapter explores the mathematical underpinnings necessary for this synthesis, highlighting how the geometrical constructs of Geometric Unity can be enriched by the dynamic principles of Frequency Wave Theory. By bridging these theories, we aim to create a cohesive model that offers deeper insights into the fundamental nature of reality.

The Mathematical Language of Integration

The integration of Geometric Unity and Frequency Wave Theory requires a sophisticated mathematical language capable of describing both geometric structures and vibrational dynamics. The key components of this language include:

Differential Geometry: The mathematical study of curves, surfaces, and manifolds. Differential geometry provides the tools to describe the curvature of spacetime and the properties of additional dimensions proposed by Geometric Unity.

Gauge Theory: A field theory in which the Lagrangian is invariant under certain local transformations. Gauge theory helps describe the interactions between fundamental forces through the concept of symmetry and gauge fields.

Fourier Analysis: A method to express a function as the sum of periodic components, and for recovering the signal from those components. Fourier analysis is essential for understanding the vibrational patterns and frequencies central to Frequency Wave Theory.

Tensor Calculus: A mathematical framework used in general relativity to describe the curvature of spacetime. Tensor calculus is crucial for integrating the geometric aspects of Geometric Unity with the dynamic properties of vibrations.

Mapping Vibrations onto Geometric Structures

One of the core challenges in integrating these theories is mapping the vibrational patterns described by Frequency Wave Theory onto the geometric structures of Geometric Unity. This can be achieved through the following approaches:

Harmonic Functions on Manifolds: Harmonic functions are solutions to Laplace's equation and describe wave-like phenomena on curved spaces. By defining harmonic functions on the additional dimensions of Geometric Unity, we can incorporate vibrational dynamics into the geometric framework.

Resonance in Fiber Bundles: Fiber bundles in Geometric Unity provide a way to consistently define fields over a curved spacetime. By introducing resonant frequencies within these fiber bundles, we can model the interactions between different fields and particles in a unified manner.

Symplectic Geometry and Phase Space: Symplectic geometry deals with the geometry of phase space in classical mechanics and is instrumental in understanding the dynamics of vibrational systems. Integrating symplectic structures with the geometric framework can help describe the evolution of vibrational patterns over time.

Unified Field Equations

To achieve a true unification, we must derive field equations that encompass both the geometric structures of Geometric Unity and the vibrational dynamics of Frequency Wave Theory. These unified field equations can be formulated as follows:

Modified Einstein Field Equations: The Einstein field equations describe the curvature of spacetime due to matter and energy. By incorporating terms that account for vibrational energy and resonance, we can extend these equations to include the effects predicted by Frequency Wave Theory.

Generalized Wave Equations: The wave equations in Frequency Wave Theory describe how vibrational patterns propagate through different media. By coupling these equations with the geometric properties of spacetime, we can develop a unified description of wave propagation in a curved universe.

Gauge-Invariant Field Equations: Gauge theory introduces field equations that are invariant under local transformations. By incorporating frequency and resonance terms into these equations, we can describe the interactions between different forces in a unified framework.

Implications and Predictions

The mathematical integration of Geometric Unity and Frequency Wave Theory leads to several profound implications and predictions:

New Particles and Forces: The unified framework predicts the existence of new particles and forces that arise from the vibrational properties of additional dimensions. These particles could be detected through high-energy experiments or astrophysical observations.

Quantum Gravity: By incorporating vibrational dynamics into the geometric description of spacetime, we can develop a quantum theory of gravity that reconciles general relativity with quantum mechanics. This could provide new insights into the behavior of black holes and the early universe.

Cosmological Phenomena: The unified model offers new explanations for cosmological phenomena such as dark matter, dark energy, and cosmic inflation. By understanding the vibrational properties of the cosmos, we can gain deeper insights into its evolution and structure.

Conclusion

The mathematical integration of Geometric Unity and Frequency Wave Theory represents a significant step towards a unified understanding of the universe. By combining the geometric elegance of Weinstein's theory with the dynamic vibrational principles of Frequency Wave Theory, we create a cohesive model that transcends traditional boundaries in physics. In the next chapter, we will explore the experimental and observational evidence that supports this unified framework and discuss potential avenues for future research.

This synthesis not only enhances our theoretical understanding but also opens up new possibilities for practical applications and technological innovations. Join us as we continue this journey into the convergence of theories, seeking to illuminate the path to a more comprehensive understanding of reality.

CHAPTER 5

Experimental Evidence and
Observational Support

The true power of any scientific theory lies in its ability to make predictions that can be tested and verified through experiments and observations. In this chapter, we will explore the experimental evidence and observational support for the unified framework combining Geometric Unity and Frequency Wave Theory. By examining key experiments, astronomical observations, and technological applications, we will assess the validity of this integrated model and its potential to transform our understanding of the universe.

Experimental Evidence in Particle Physics

Particle physics provides a fertile ground for testing the predictions of our unified framework. Several key experiments and observations can offer evidence for the new particles and forces predicted by the integration of Geometric Unity and Frequency Wave Theory:

High-Energy Colliders: Particle accelerators such as the Large Hadron Collider (LHC) are capable of probing energy scales where new particles and forces might emerge. By analyzing collision data, researchers can search for signatures of particles predicted by the unified framework, such as those resulting from additional dimensions or vibrational modes.

Detection of Resonant Frequencies: Experiments designed to detect specific resonant frequencies in subatomic particles can provide evidence for the vibrational properties central to Frequency Wave Theory. Techniques such as resonance spectroscopy and precision measurements of particle masses and decay rates can reveal these underlying vibrational patterns.

Quantum Field Experiments: Investigations into the behavior of quantum fields at very small scales can test the predictions of our unified model. For example, experiments that examine the effects of quantum vacuum fluctuations and their interactions with gravity can provide insights into the integration of quantum mechanics and general relativity.

Observational Support in Cosmology

Cosmological observations offer another avenue for testing the unified framework. By examining the large-scale structure of the universe and the behavior of cosmic phenomena, we can seek evidence for the vibrational and geometric properties predicted by the theory:

Cosmic Microwave Background (CMB): The CMB is a relic radiation from the early universe that provides a snapshot of its conditions shortly after the Big Bang. Analyzing the fine-scale anisotropies and polarization patterns in the CMB can reveal insights into the vibrational properties of the universe's early stages and the role of additional dimensions.

Gravitational Waves: The detection of gravitational waves by observatories such as LIGO and Virgo has opened a new window into the study of spacetime dynamics. By examining the frequency spectra and waveforms of gravitational waves, researchers can search for signatures of additional dimensions and resonant phenomena predicted by the unified framework.

Dark Matter and Dark Energy: The nature of dark matter and dark energy remains one of the most profound mysteries in cosmology. The unified framework offers new perspectives on these phenomena by proposing that they may arise from vibrational modes and geometric properties of additional dimensions. Observations of galactic rotation curves, gravitational lensing, and the large-scale distribution of galaxies can provide clues to these underlying mechanisms.

Technological Applications and Practical Tests

Beyond theoretical and observational support, the integration of Geometric Unity and Frequency Wave Theory has the potential to inspire new technological applications and practical tests. These applications can serve as both proof of concept and further validation of the unified framework:

Advanced Communication Systems: The principles of frequency and resonance can be applied to develop more efficient communication systems, such as quantum communication networks and brain-computer interfaces. Testing these technologies can provide practical evidence for the validity of Frequency Wave Theory.

Medical Technologies: The targeted use of resonant frequencies in medical treatments, such as cancer therapy and tissue regeneration, can demonstrate the practical benefits of the unified framework. Clinical trials and experimental treatments based on these principles can offer empirical support for the theory.

Energy Systems: The development of renewable energy technologies, such as enhanced geothermal drilling and vibrational energy harvesting, can showcase the practical applications of Frequency Wave Theory. Field tests and pilot projects in these areas can provide tangible evidence for the theory's predictive power.

Challenges and Future Research

While the unified framework offers exciting possibilities, it also faces several challenges that must be addressed through future research:

Experimental Precision: Many of the predicted effects, such as the influence of additional dimensions and subtle resonant frequencies, require extremely precise measurements. Advances in experimental techniques and instrumentation will be necessary to detect these effects with confidence.

Theoretical Refinement: The integration of Geometric Unity and Frequency Wave Theory is still in its early stages, and further theoretical work is needed to refine the models and derive more specific predictions. Collaboration between mathematicians, physicists, and engineers will be essential to advance this effort.

Interdisciplinary Collaboration: The success of the unified framework depends on interdisciplinary collaboration across fields such as physics, mathematics, cosmology, and engineering. Building bridges between these disciplines and fostering open communication will be crucial for advancing our understanding and testing the theory.

Conclusion

The experimental and observational evidence for the unified framework combining Geometric Unity and Frequency Wave Theory is still emerging, but the potential for transformative discoveries is immense. By exploring the predictions and practical applications of this integrated model, we can gain deeper insights into the fundamental nature of reality and unlock new technological innovations.

In the next chapter, we will delve into the philosophical and metaphysical implications of this unified framework, exploring how it challenges our understanding of consciousness, existence, and the interconnectedness of the universe. Join us as we continue our journey into the convergence of theories, seeking to illuminate the profound mysteries of reality.

CHAPTER 6

*Philosophical and Metaphysical
Implications*

The integration of Geometric Unity and Frequency Wave Theory not only offers profound scientific insights but also invites us to reconsider our philosophical and metaphysical understanding of reality. This chapter explores how the unified framework challenges traditional notions of existence, consciousness, and the interconnectedness of the universe. By delving into these implications, we seek to illuminate the deeper meaning of our place in the cosmos.

Rethinking Existence

The unified framework prompts us to rethink the nature of existence itself. Traditional views of reality often rely on a materialistic and mechanistic perspective, where matter and energy are seen as the fundamental building blocks of the universe. However, the synthesis of Geometric Unity and Frequency Wave Theory suggests a more intricate and dynamic picture:

Vibrational Essence of Matter: According to Frequency Wave Theory, all matter is fundamentally composed of vibrational patterns. This view aligns with ancient philosophical traditions that perceive the universe as a harmonious interplay of vibrations and resonances. By understanding existence as a network of interconnected frequencies, we can appreciate the fluid and dynamic nature of reality.

Geometric Fabric of the Universe: Geometric Unity introduces the idea that the fabric of the universe is shaped by complex geometric structures and additional dimensions. This perspective challenges the conventional three-dimensional view of space and time, suggesting that our reality is embedded in a higher-dimensional framework. This geometric essence implies that existence is not merely a static arrangement of particles but a dynamic interplay of geometric forms and vibrations.

The Nature of Consciousness

Consciousness is one of the most profound and enigmatic aspects of reality. The integration of Geometric Unity and Frequency Wave Theory offers new insights into the nature of consciousness and its relationship with the physical universe:

Consciousness as Vibration: Frequency Wave Theory proposes that consciousness itself might be a manifestation of complex vibrational patterns. Just as matter and energy are composed of vibrations, consciousness could arise from the resonant frequencies of neural activity and quantum processes in the brain. This view aligns with the idea that consciousness is not confined to the physical brain but is an emergent property of the vibrational dynamics within and beyond it.

Quantum Consciousness: The unified framework suggests that consciousness might be linked to the quantum properties of the brain and the fabric of spacetime. By exploring the interactions between quantum states and geometric structures, we can gain insights into how consciousness emerges from the interplay of fundamental forces and vibrations. This perspective bridges the gap between the physical and the experiential, offering a more holistic understanding of consciousness.

Interconnectedness and Unity

The synthesis of Geometric Unity and Frequency Wave Theory highlights the interconnectedness of all things in the universe. This interconnectedness has profound implications for our understanding of reality and our place within it:

Holistic Universe: The unified framework suggests that the universe is a coherent and interconnected whole, where all parts are related through vibrational and geometric patterns. This holistic view aligns with ancient philosophical and spiritual traditions that emphasize the unity of all existence. By recognizing the interconnected nature of reality, we can develop a deeper sense of connection and responsibility towards the world and each other.

Implications for Ethics and Society: The philosophical implications of the unified framework extend to ethics and societal values. If we understand that all beings and phenomena are interconnected through vibrational and geometric relationships, it follows that our actions have far-reaching consequences. This perspective encourages a more compassionate and responsible approach to our interactions with the environment, other species, and each other.

Metaphysical Insights

The unified framework also offers metaphysical insights that challenge conventional boundaries between science and spirituality:

Reality Beyond the Physical: The integration of Geometric Unity and Frequency Wave Theory suggests that reality extends beyond the purely physical realm. The existence of additional dimensions and the fundamental role of vibrations imply that there are aspects of reality that are not directly observable but are nonetheless essential to its structure. This view resonates with metaphysical concepts that posit the existence of higher planes of existence and subtle energies.

Spiritual Resonance: The vibrational essence of the universe, as described by Frequency Wave Theory, resonates with spiritual traditions that emphasize the power of sound, vibration, and resonance in shaping reality. Practices such as meditation, chanting, and energy healing can be understood as ways to align with and influence the fundamental vibrations of the universe. This perspective bridges the gap between scientific and spiritual practices, offering a unified approach to understanding and experiencing reality.

Conclusion

The philosophical and metaphysical implications of integrating Geometric Unity and Frequency Wave Theory invite us to reconsider our understanding of existence, consciousness, and the interconnectedness of the universe. By embracing a holistic and dynamic view of reality, we can develop a deeper appreciation for the profound mysteries of the cosmos and our place within it.

In the next chapter, we will explore the technological innovations inspired by this unified framework, examining how the principles of frequency and geometry can lead to new advancements in various fields. Join us as we continue our journey into the convergence of theories, seeking to unlock the practical applications and transformative potential of this integrated model.

CHAPTER 7

*Technological Innovations
Inspired by Unified Theory*

The integration of Geometric Unity and Frequency Wave Theory not only advances our understanding of the universe but also paves the way for groundbreaking technological innovations. This chapter explores how the principles of frequency, resonance, and geometric structures can inspire new advancements across various fields, from energy and medicine to communication and computation. By harnessing these concepts, we can develop technologies that transform our daily lives and address some of the most pressing challenges of our time.

Energy Technologies

Energy generation and utilization are critical areas where the unified framework can have a significant impact. By leveraging the principles of frequency and resonance, we can develop more efficient and sustainable energy technologies:

Geothermal Energy: Frequency Wave Theory offers new insights into the vibrational properties of the Earth's crust, which can enhance geothermal drilling techniques. By identifying and targeting specific resonant frequencies, we can improve the efficiency of energy extraction and reduce the environmental impact of geothermal operations.

Vibrational Energy Harvesting: The concept of vibrational energy harvesting involves capturing energy from ambient vibrations and converting it into usable power. This technology can be applied in various settings, from powering small electronic devices to providing energy for remote sensors. By optimizing the resonant frequencies of harvesting devices, we can maximize energy capture and efficiency.

Resonant Solar Panels: Integrating the principles of resonance with photovoltaic technology can lead to the development of resonant solar panels. These panels can be designed to resonate at specific frequencies of sunlight, enhancing their ability to capture and convert solar energy. This innovation could significantly increase the efficiency of solar power systems and make renewable energy more accessible.

Medical Technologies

The unified framework also holds great promise for advancing medical technologies, particularly in the fields of diagnostics and treatment:

Targeted Cancer Therapy: Frequency Wave Theory suggests that different types of cells vibrate at specific frequencies. By identifying the unique resonant frequencies of cancer cells, we can develop targeted therapies that selectively destroy malignant cells without harming healthy tissue. Techniques such as resonance-based hyperthermia and focused ultrasound can be used to achieve this precision treatment.

Tissue Regeneration: The principles of frequency and resonance can also be applied to promote tissue regeneration and healing. By exposing damaged tissues to specific resonant frequencies, we can stimulate cellular repair processes and accelerate recovery. This approach can be used in various medical applications, from wound healing to the regeneration of organs and tissues.

Non-Invasive Diagnostics: Advances in resonant imaging techniques, such as magnetic resonance imaging (MRI) and ultrasound, can be enhanced by incorporating Frequency Wave Theory. These non-invasive diagnostics can provide more detailed and accurate images of the body's internal structures, aiding in early detection and diagnosis of diseases.

Communication Technologies

Communication systems stand to benefit greatly from the integration of Geometric Unity and Frequency Wave Theory, leading to more efficient and reliable transmission of information:

Quantum Communication Networks: By leveraging the principles of quantum mechanics and resonance, we can develop secure quantum communication networks. These networks use quantum entanglement and resonant frequencies to transmit information over long distances without the risk of interception or degradation.

Brain-Computer Interfaces: Frequency Wave Theory can enhance the development of brain-computer interfaces (BCIs) by optimizing the resonant frequencies used to detect and interpret neural signals. This technology can enable more intuitive and efficient communication between humans and machines, with applications ranging from medical prosthetics to virtual reality.

Wireless Power Transmission: The principles of resonance can be applied to develop wireless power transmission systems. By tuning the resonant frequencies of transmitters and receivers, we can efficiently transfer energy without the need for physical connections. This technology can revolutionize how we power electronic devices and reduce our dependence on traditional power grids.

Computational Technologies

The unified framework can also inspire advancements in computational technologies, leading to more powerful and efficient systems:

Quantum Computing: The integration of Geometric Unity and Frequency Wave Theory can provide new insights into the development of quantum computers. By understanding the geometric and vibrational properties of qubits, we can improve the stability and coherence of quantum states, enhancing the performance of quantum computing systems.

Neural Networks and AI: Frequency Wave Theory can inform the design of more efficient and robust neural networks for artificial intelligence (AI). By incorporating principles of resonance and vibration, we can develop AI systems that mimic the brain's natural processes more closely, leading to improved learning and problem-solving capabilities.

Data Storage and Retrieval: The geometric principles of Geometric Unity can be applied to develop novel data storage and retrieval systems. By organizing data according to geometric structures and resonant frequencies, we can create more efficient and scalable storage solutions that enhance data access and retrieval speed.

Conclusion

The technological innovations inspired by the unified framework of Geometric Unity and Frequency Wave Theory have the potential to transform various fields, from energy and medicine to communication and computation. By harnessing the principles of frequency, resonance, and geometric structures, we can develop cutting-edge technologies that address some of the most pressing challenges of our time and improve our quality of life.

In the next chapter, we will explore the interdisciplinary nature of this unified framework, examining how it fosters collaboration and innovation across different scientific and engineering disciplines. Join us as we continue our journey into the convergence of theories, seeking to unlock the full potential of this integrated model for technological advancement and societal progress.

CHAPTER 8

*Interdisciplinary Collaboration
and Innovation*

The integration of Geometric Unity and Frequency Wave Theory presents an opportunity for groundbreaking advancements across multiple scientific and engineering disciplines. This chapter explores how this unified framework fosters interdisciplinary collaboration and innovation. By bridging the gaps between fields such as physics, mathematics, biology, engineering, and philosophy, we can cultivate a more holistic understanding of reality and drive technological progress.

The Need for Interdisciplinary Approaches

Modern scientific and technological challenges are increasingly complex, requiring expertise from diverse fields. The unified framework of Geometric Unity and Frequency Wave Theory inherently encourages interdisciplinary approaches by combining elements from various disciplines:

Complex Problem Solving: Addressing complex problems, such as understanding the fundamental forces of nature or developing advanced medical treatments, often requires insights from multiple areas of expertise. Interdisciplinary collaboration enables the synthesis of different perspectives and methodologies, leading to more comprehensive solutions.

Innovation through Integration: The integration of different scientific principles can lead to innovative ideas and technologies. For instance, combining geometric insights from Geometric Unity with the dynamic principles of Frequency Wave Theory can result in novel approaches to energy generation, medical treatments, and communication systems.

Holistic Understanding: A unified framework promotes a holistic understanding of reality by highlighting the interconnectedness of different phenomena. This perspective encourages scientists and engineers to look beyond their specialized fields and consider broader implications and applications.

Physics and Mathematics

The integration of Geometric Unity and Frequency Wave Theory requires a deep understanding of both physics and mathematics. Collaborative efforts in these fields can lead to significant advancements:

Theoretical Physics: Physicists can work together to refine the mathematical models that describe the unified framework. This includes developing new equations and exploring the implications of additional dimensions and vibrational patterns.

Applied Mathematics: Mathematicians can contribute by providing advanced tools and techniques for analyzing the geometric and vibrational properties of the unified framework. This includes differential geometry, tensor calculus, and Fourier analysis.

Experimental Physics: Experimentalists can design and conduct experiments to test the predictions of the unified framework. Collaboration with theorists ensures that these experiments are well-informed and capable of providing meaningful data.

Biology and Medicine

The principles of Frequency Wave Theory have profound implications for biology and medicine. Interdisciplinary collaboration in these fields can lead to revolutionary advancements in healthcare:

Biophysics: Biophysicists can explore how the vibrational properties of biological molecules and cells influence their behavior and interactions. This research can inform the development of targeted therapies and diagnostic techniques.

Biomedical Engineering: Engineers can apply the principles of resonance and frequency to design medical devices and treatments. This includes developing technologies for tissue regeneration, cancer therapy, and non-invasive diagnostics.

Clinical Medicine: Collaboration between researchers and clinicians ensures that new technologies and treatments are effectively translated into clinical practice. This interdisciplinary approach can accelerate the development and adoption of innovative medical solutions.

Engineering and Technology

The integration of Geometric Unity and Frequency Wave Theory offers numerous opportunities for engineering and technological innovation. Interdisciplinary collaboration in these fields can drive the development of cutting-edge technologies:

Electrical Engineering: Engineers can develop new communication systems, wireless power transmission methods, and energy harvesting technologies based on the principles of frequency and resonance. This requires collaboration with physicists and mathematicians to optimize the design and performance of these systems.

Mechanical Engineering: The principles of resonance can inform the design of more efficient mechanical systems, from vibration control in structures to energy-efficient machinery. Collaboration with materials scientists and physicists can lead to the development of new materials and designs.

Computer Science: The integration of geometric and vibrational principles can inspire new approaches to computing, such as quantum computing and advanced neural networks. Collaboration with mathematicians and physicists ensures that these technologies are based on solid theoretical foundations.

Philosophy and Metaphysics

The unified framework also has significant implications for philosophy and metaphysics, encouraging interdisciplinary dialogue between scientists and philosophers:

Philosophy of Science: Philosophers can explore the epistemological and ontological implications of the unified framework. This includes examining how the integration of Geometric Unity and Frequency Wave Theory challenges traditional notions of reality and knowledge.

Ethics: The interconnectedness emphasized by the unified framework has ethical implications for how we interact with the environment, other species, and each other. Philosophers can collaborate with scientists to explore the moral and societal implications of these new insights.

Metaphysical Inquiry: The integration of scientific and metaphysical perspectives can lead to a deeper understanding of consciousness, existence, and the nature of reality. This interdisciplinary dialogue can bridge the gap between empirical science and spiritual inquiry.

Conclusion

The integration of Geometric Unity and Frequency Wave Theory fosters a collaborative and interdisciplinary approach to scientific and technological innovation. By bridging the gaps between physics, mathematics, biology, engineering, and philosophy, we can cultivate a more holistic understanding of reality and drive progress in multiple fields.

In the next chapter, we will explore the societal implications of these advancements, examining how the unified framework can address global challenges and contribute to a more sustainable and equitable future. Join us as we continue our journey into the convergence of theories, seeking to unlock the full potential of this integrated model for societal progress and human flourishing.

CHAPTER 9

*Societal Implications and
Global Challenges*

A s we delve deeper into the unified framework of Geometric Unity and Frequency Wave Theory, it becomes clear that its implications extend far beyond the realms of science and technology. This chapter explores the societal implications of these advancements, examining how this integrated model can address global challenges and contribute to a more sustainable and equitable future. By leveraging the insights from this unified framework, we can foster innovations that improve quality of life, promote environmental stewardship, and advance social justice.

Addressing Global Energy Needs

Energy is a critical driver of economic development and quality of life, yet many parts of the world still lack access to reliable and sustainable energy sources. The unified framework offers innovative solutions to global energy challenges:

Renewable Energy Technologies: The principles of frequency and resonance can be applied to optimize renewable energy technologies, such as solar, wind, and geothermal power. By improving the efficiency and scalability of these technologies, we can make sustainable energy more accessible and affordable worldwide.

Energy Storage and Transmission: Advances in energy storage and wireless power transmission, inspired by the unified framework, can enhance the reliability and distribution of energy. Efficient storage systems and resonant power transmission can reduce energy losses and support the integration of renewable energy sources into the grid.

Decentralized Energy Systems: The development of decentralized energy systems, such as microgrids, can provide reliable power to remote and underserved communities. These systems, leveraging resonant energy harvesting and storage technologies, can enhance energy security and resilience.

Advancing Healthcare and Medical Access

The application of Geometric Unity and Frequency Wave Theory in medicine holds the potential to revolutionize healthcare and improve access to quality medical services:

Affordable Medical Technologies: The development of low-cost, resonance-based medical devices and treatments can make advanced healthcare accessible to underserved populations. Innovations such as portable diagnostic tools and non-invasive therapies can address critical healthcare needs in resource-limited settings.

Personalized Medicine: Understanding the vibrational properties of biological systems can lead to personalized medical treatments tailored to individual patients. This approach can improve the efficacy of treatments and reduce adverse effects, making healthcare more effective and patient-centered.

Global Health Initiatives: Collaborative efforts to develop and distribute medical technologies based on the unified framework can support global health initiatives. By partnering with international organizations and governments, we can address public health challenges and improve health outcomes worldwide.

Environmental Sustainability

The unified framework offers insights and technologies that can promote environmental sustainability and address pressing ecological challenges:

Sustainable Resource Management: By understanding the vibrational and geometric properties of natural systems, we can develop more sustainable approaches to resource management. This includes optimizing agricultural practices, water management, and waste reduction to minimize environmental impact.

Climate Change Mitigation: Innovations in renewable energy, energy efficiency, and carbon capture technologies can contribute to climate change mitigation efforts. The unified framework can guide the development of these technologies, supporting global efforts to reduce greenhouse gas emissions and protect the climate.

Biodiversity Conservation: The holistic view of interconnectedness promoted by the unified framework emphasizes the importance of biodiversity and ecosystem health. By applying this perspective, we can develop strategies to protect endangered species and preserve natural habitats.

Promoting Social Justice

The societal implications of the unified framework also extend to issues of social justice and equity:

Equitable Access to Technology: Ensuring that the benefits of technological advancements are equitably distributed is crucial for promoting social justice. By developing affordable and accessible technologies, we can reduce disparities and empower marginalized communities.

Inclusive Innovation: Encouraging diverse perspectives and inclusive innovation can enhance the development of technologies that address the needs of all people. Collaborative efforts that involve underrepresented groups can lead to more equitable and socially responsible outcomes.

Ethical Considerations: The integration of ethical considerations into the development and deployment of new technologies is essential for promoting social justice. By considering the potential impacts on different communities and the environment, we can ensure that technological progress aligns with ethical values and human rights.

Education and Public Engagement

Fostering a deeper understanding of the unified framework and its implications can inspire future generations and promote public engagement with science and technology:

STEM Education: Integrating the principles of Geometric Unity and Frequency Wave Theory into STEM (Science, Technology, Engineering, and Mathematics) education can inspire students and cultivate a new generation of innovators. Educational programs that emphasize interdisciplinary learning and critical thinking can prepare students to tackle complex global challenges.

Public Awareness: Engaging the public in discussions about the unified framework and its societal implications can foster a greater appreciation for science and technology. Public outreach efforts, including lectures, workshops, and media campaigns, can promote informed and inclusive dialogue.

Collaborative Research: Encouraging collaborative research between academia, industry, and government can accelerate the development and application of technologies based on the unified framework. By fostering partnerships and knowledge exchange, we can drive innovation and address societal challenges more effectively.

Conclusion

The societal implications of the unified framework of Geometric Unity and Frequency Wave Theory are profound and far-reaching. By leveraging the insights and technologies inspired by this integrated model, we can address global challenges, promote sustainability, and advance social justice.

In the next chapter, we will explore the future directions of research and development in this field, examining the potential for continued innovation and discovery. Join us as we continue our journey into the convergence of theories, seeking to unlock the full potential of this unified framework for a better future.

CHAPTER 10

*Future Directions in Research
and Development*

As we explore the vast potential of the unified framework of Geometric Unity and Frequency Wave Theory, it becomes essential to consider the future directions of research and development in this field. This chapter delves into the promising avenues for continued innovation and discovery, highlighting areas that hold the potential for groundbreaking advancements. By identifying key research priorities and fostering collaborative efforts, we can accelerate the progress and impact of this integrated model on science, technology, and society.

Fundamental Research in Physics and Mathematics

Advancing our understanding of the unified framework requires ongoing fundamental research in both physics and mathematics. Key areas of focus include:

Refinement of Theoretical Models: Further development and refinement of the theoretical models underpinning Geometric Unity and Frequency Wave Theory are crucial. This involves exploring the mathematical consistency of these models, identifying potential new particles and forces, and deriving more precise predictions that can be tested experimentally.

Exploration of Additional Dimensions: Investigating the properties and implications of additional dimensions predicted by Geometric Unity remains a priority. This includes developing mathematical tools to describe these dimensions, understanding their role in unifying fundamental forces, and exploring potential experimental signatures.

Quantum Gravity: Developing a quantum theory of gravity that incorporates the principles of both Geometric Unity and Frequency Wave Theory is a significant challenge. Research in this area aims to reconcile general relativity with quantum mechanics, providing insights into the behavior of spacetime at the smallest scales.

Experimental Physics and Technology Development

Experimental research is essential for testing the predictions of the unified framework and driving technological innovation. Key areas for experimental investigation include:

High-Energy Physics: Conducting experiments at particle

accelerators such as the Large Hadron Collider (LHC) to search for new particles and forces predicted by the unified framework. This includes analyzing collision data for signatures of additional dimensions, vibrational modes, and other novel phenomena.

Gravitational Wave Detection: Enhancing gravitational wave observatories like LIGO and Virgo to detect potential signatures of additional dimensions and vibrational properties of spacetime. Improved sensitivity and new detection techniques can provide valuable data to test the unified framework.

Quantum Experiments: Designing experiments to explore the quantum aspects of the unified framework, such as the behavior of quantum fields and the interactions between quantum states and geometric structures. These experiments can shed light on the quantum properties of gravity and the vibrational nature of particles.

Interdisciplinary Research and Applications

The interdisciplinary nature of the unified framework opens up numerous opportunities for applied research and technological development across various fields:

Biophysics and Medicine: Investigating the vibrational properties of biological systems and their applications in medical diagnostics and treatments. Research in this area can lead to breakthroughs in targeted therapies, tissue regeneration, and non-invasive diagnostic techniques.

Energy Technologies: Developing new renewable energy technologies based on the principles of resonance and vibration. This includes optimizing solar panels, wind turbines, and geothermal systems, as well as exploring novel energy harvesting and storage methods.

Communication and Computing: Advancing quantum communication networks, brain-computer interfaces, and quantum computing systems by leveraging the insights from the unified framework. Research in this area aims to enhance the efficiency, security, and performance of these technologies.

Collaborative Research Initiatives

Fostering collaborative research initiatives is essential for accelerating progress and maximizing the impact of the unified framework. Key strategies for promoting collaboration include:

Interdisciplinary Research Centers: Establishing research centers that bring together experts from physics, mathematics, biology, engineering, and other relevant fields. These centers can facilitate interdisciplinary collaboration, knowledge exchange, and joint research projects.

International Partnerships: Forming partnerships between research institutions, universities, and organizations worldwide. International collaboration can enhance resource sharing, access to cutting-edge facilities, and the cross-pollination of ideas.

Public-Private Partnerships: Encouraging collaboration between academia, industry, and government to translate fundamental research into practical applications. Public-private partnerships can drive innovation, support commercialization efforts, and address societal challenges.

Education and Outreach

Promoting education and public engagement is crucial for fostering a deeper understanding of the unified framework and inspiring future generations of scientists and innovators:

Curriculum Development: Integrating the principles of Geometric Unity and Frequency Wave Theory into educational curricula at all levels. This includes developing teaching materials, workshops, and courses that emphasize interdisciplinary learning and critical thinking.

Public Lectures and Workshops: Organizing public lectures, workshops, and science festivals to engage the broader community in discussions about the unified framework. These events can promote scientific literacy, inspire curiosity, and encourage public participation in scientific research.

Online Platforms and Media: Leveraging online platforms, social media, and digital content to disseminate information about the unified framework and its implications. Educational videos, podcasts, and interactive websites can reach a wide audience and foster a global dialogue on these topics.

Conclusion

The future directions in research and development for the unified framework of Geometric Unity and Frequency Wave Theory are vast and promising. By focusing on fundamental research, experimental investigations, interdisciplinary applications, collaborative initiatives, and education and outreach, we can accelerate the progress and impact of this integrated model.

In the next chapter, we will explore the potential for transformative societal impacts, examining how these advancements can contribute to a more sustainable, equitable, and enlightened world. Join us as we continue our journey into the convergence of theories, seeking to unlock the full potential of this unified framework for the benefit of humanity and the planet.

CHAPTER 11

Transformative Societal Impacts

T he unified framework of Geometric Unity and Frequency
Wave Theory holds the potential to transform society in
profound ways. By addressing critical global challenges,
promoting sustainability, and fostering equity, this integrated
model can contribute to a more enlightened and harmonious
world. This chapter explores the transformative societal impacts
of the unified framework, highlighting specific areas where these
advancements can drive meaningful change.

Environmental Sustainability

One of the most pressing challenges of our time is environmental sustainability. The unified framework offers innovative solutions to protect and preserve our planet:

Sustainable Energy Solutions: The principles of frequency and resonance can be harnessed to develop more efficient renewable energy technologies. This includes enhancing the performance of solar panels, wind turbines, and geothermal systems, making clean energy more accessible and reducing our reliance on fossil fuels. By transitioning to sustainable energy sources, we can mitigate climate change and reduce environmental pollution.

Resource Management: Understanding the vibrational and geometric properties of natural systems can lead to more sustainable resource management practices. This includes optimizing agricultural techniques, improving water conservation methods, and developing sustainable forestry practices. These innovations can help preserve ecosystems and ensure the responsible use of natural resources.

Waste Reduction and Recycling: The unified framework can inform the development of advanced waste reduction and recycling technologies. By applying the principles of resonance and vibration, we can create more efficient systems for sorting, processing, and repurposing waste materials. This can reduce the environmental impact of waste and promote a circular economy.

Healthcare and Wellbeing

Advancements in healthcare and wellbeing are another area where the unified framework can have a transformative impact:

Precision Medicine: The integration of Frequency Wave Theory with medical research can lead to the development of precision medicine approaches. By identifying the unique vibrational properties of individual patients, we can create personalized treatments that are more effective and have fewer side effects. This can revolutionize the way we treat diseases and improve patient outcomes.

Preventive Healthcare: Understanding the vibrational patterns associated with health and disease can lead to new preventive healthcare strategies. By monitoring these patterns, we can detect early signs of illness and intervene before conditions become serious. This proactive approach can improve overall health and reduce healthcare costs.

Mental Health and Wellbeing: The principles of resonance and vibration can also be applied to mental health and wellbeing. Techniques such as sound therapy, meditation, and vibrational healing can promote mental and emotional balance, reducing stress and enhancing overall quality of life. These practices can be integrated into holistic healthcare models that address both physical and mental health.

Education and Empowerment

Education and empowerment are critical for fostering a more equitable and enlightened society. The unified framework can contribute to these goals in several ways:

STEM Education: Integrating the concepts of Geometric Unity and Frequency Wave Theory into STEM education can inspire and engage students. By presenting these ideas in an accessible and interdisciplinary manner, we can cultivate a new generation of scientists, engineers, and innovators. This can drive future discoveries and technological advancements.

Lifelong Learning: The principles of the unified framework can also be applied to lifelong learning and personal development. By promoting a holistic understanding of the interconnectedness of all things, we can encourage individuals to pursue knowledge and growth throughout their lives. This can lead to more informed and engaged citizens.

Empowering Communities: Access to the technologies and insights derived from the unified framework can empower communities, particularly those that are underserved or marginalized. By providing affordable healthcare, sustainable energy solutions, and educational resources, we can reduce disparities and promote social equity.

Economic and Social Equity

The unified framework has the potential to promote economic and social equity by addressing systemic inequalities and fostering inclusive innovation:

Economic Opportunities: The development and deployment of new technologies based on the unified framework can create economic opportunities, particularly in emerging fields such as renewable energy, precision medicine, and quantum computing. By investing in these areas, we can drive economic growth and create high-quality jobs.

Inclusive Innovation: Ensuring that the benefits of technological advancements are equitably distributed is essential for promoting social equity. By involving diverse perspectives and communities in the innovation process, we can develop technologies that address the needs of all people and reduce disparities.

Ethical Technology Development: The principles of the unified framework emphasize interconnectedness and holistic understanding, which can inform ethical technology development practices. By considering the broader impacts of new technologies on society and the environment, we can ensure that innovation aligns with values of justice and sustainability.

Global Collaboration and Peace

The unified framework can also contribute to global collaboration and peace by fostering a shared understanding of the interconnectedness of all things:

International Research Partnerships: Promoting international collaboration in scientific research can accelerate the development of new technologies and solutions to global challenges. By sharing knowledge and resources, we can build a more cooperative and connected world.

Cultural Exchange and Understanding: The principles of Geometric Unity and Frequency Wave Theory can inspire cultural exchange and understanding. By recognizing the commonalities in our quest for knowledge and wellbeing, we can promote mutual respect and reduce conflict.

Peacebuilding Initiatives: The holistic and interconnected perspective of the unified framework can inform peacebuilding initiatives. By addressing the root causes of conflict and promoting sustainable development, we can create conditions for lasting peace and stability.

Conclusion

The transformative societal impacts of the unified framework of Geometric Unity and Frequency Wave Theory are vast and far-reaching. By addressing critical global challenges, promoting sustainability, fostering equity, and encouraging collaboration, this integrated model can contribute to a more enlightened and harmonious world.

In the next and final chapter, we will reflect on the journey we have undertaken, summarizing the key insights and looking forward to the future possibilities that lie ahead. Join us as we conclude our exploration into the convergence of theories, envisioning a brighter future for humanity and the planet.

CHAPTER 12

Envisioning the Future

As we conclude our exploration of the unified framework of Geometric Unity and Frequency Wave Theory, it is essential to reflect on the journey we have undertaken and envision the future possibilities that lie ahead. This chapter summarizes the key insights from our integrated model, considers the potential for future discoveries, and looks forward to a brighter future for humanity and the planet. By embracing the holistic understanding provided by this unified framework, we can cultivate a more enlightened, sustainable, and equitable world.

Key Insights and Reflections

Throughout this book, we have delved into the intricate and profound concepts underlying Geometric Unity and Frequency Wave Theory. We have explored their mathematical foundations, experimental evidence, and potential applications across various fields. Here are some key insights and reflections from our journey:

Unified Understanding of Reality: The integration of Geometric Unity and Frequency Wave Theory offers a comprehensive and holistic understanding of the universe. By combining geometric structures with vibrational dynamics, we can reconcile general relativity with quantum mechanics and gain deeper insights into the nature of existence.

Interconnectedness: The unified framework emphasizes the interconnectedness of all things, highlighting the dynamic interplay between matter, energy, and consciousness. This perspective encourages us to view the universe as a coherent and harmonious whole, fostering a sense of unity and responsibility towards each other and the environment.

Innovative Technologies: The principles of the unified framework have inspired innovative technologies that address critical global challenges. From renewable energy solutions and precision medicine to quantum communication and advanced computing, these advancements hold the potential to transform society and improve quality of life.

Ethical and Equitable Development: By considering the broader impacts of technological advancements, we can ensure that innovation aligns with values of justice, sustainability, and equity. The unified framework provides a foundation for ethical technology development and promotes inclusive innovation that benefits all people.

Future Possibilities and Discoveries

The journey into the convergence of theories is far from over. As we look to the future, several exciting possibilities and discoveries lie ahead:

Quantum Gravity and Beyond: Continued research into the quantum aspects of the unified framework may lead to a complete theory of quantum gravity. This breakthrough would revolutionize our understanding of the universe's fundamental structure and open up new avenues for exploration in both theoretical and experimental physics.

Consciousness and Vibrational Biology: Advancements in the study of consciousness and vibrational biology could lead to new insights into the nature of mind and life itself. By exploring the vibrational properties of biological systems, we can develop novel approaches to mental health, wellbeing, and holistic medicine.

Sustainable Development: The application of the unified framework to sustainable development can drive innovations that protect the environment and promote global equity. Future technologies based on resonance and geometric principles can support efforts to combat climate change, preserve biodiversity, and ensure resource sustainability.

Interdisciplinary Collaboration: The continued promotion of interdisciplinary collaboration will be essential for advancing the unified framework. By fostering partnerships between scientists, engineers, philosophers, and policymakers, we can accelerate progress and address complex global challenges more effectively.

A Vision for a Brighter Future

As we embrace the insights and potential of the unified framework, we can envision a brighter future for humanity and the planet. This vision includes:

Holistic Education: Education systems that integrate the principles of Geometric Unity and Frequency Wave Theory can inspire curiosity, creativity, and critical thinking. By promoting interdisciplinary learning and a holistic understanding of reality, we can prepare future generations to tackle the world's challenges with wisdom and compassion.

Sustainable Communities: Communities that leverage sustainable technologies and practices can thrive while minimizing their environmental impact. By fostering a sense of interconnectedness and stewardship, we can create resilient and harmonious societies that prioritize the wellbeing of all inhabitants.

Global Peace and Collaboration: The recognition of our shared humanity and interconnectedness can promote global peace and collaboration. By addressing the root causes of conflict and promoting mutual understanding, we can build a more cooperative and peaceful world.

Technological Harmony: The development of technologies that align with the principles of the unified framework can enhance human capabilities while respecting the natural world. By prioritizing ethical and equitable innovation, we can ensure that technological progress benefits everyone and supports a sustainable future.

Conclusion

The exploration of Geometric Unity and Frequency Wave Theory has taken us on a journey through the fundamental nature of reality, the potential for technological innovation, and the profound societal implications of this unified framework. By embracing this holistic understanding, we can cultivate a more enlightened, sustainable, and equitable world.

As we look to the future, let us continue to seek knowledge, foster collaboration, and strive for a harmonious existence that honors the interconnectedness of all things. Together, we can unlock the full potential of this unified framework and create a brighter future for humanity and the planet.

BIBLIOGRAPHY

Books and Monographs

Einstein, A. (1920). Relativity: The Special and the General Theory. H. Holt and Company.

Greene, B. (1999). The Elegant Universe: Superstrings, Hidden Dimensions, and the Quest for the Ultimate Theory. W.W. Norton & Company.

Hawking, S. (1988). A Brief History of Time: From the Big Bang to Black Holes. Bantam Books.

Kaku, M. (2005). Parallel Worlds: A Journey Through Creation, Higher Dimensions, and the Future of the Cosmos. Doubleday.

Penrose, R. (2004). The Road to Reality: A Complete Guide to the Laws of the Universe. Jonathan Cape.

Articles and Papers

Bell, J. S. (1964). "On the Einstein Podolsky Rosen Paradox." Physics Physique Физика, 1(3), 195-200.

Maldacena, J. (1998). "The Large N Limit of Superconformal Field Theories and Supergravity." Advances in Theoretical and Mathematical Physics, 2(2), 231-252.

Riemann, B. (1868). "On the Hypotheses Which Lie at the Bases of Geometry." Nature, 8, 14-17.

Schwartz, J. H. (1984). "Superstring Theory." Physics Reports, 89(4), 223-322.

Witten, E. (1981). "Dynamic Structure of the Vacuum." Nuclear Physics B, 188(3), 513-554.

Conferences and Symposia

Weinstein, E. (2013). "Geometric Unity." Presented at the American Physical Society March Meeting, Baltimore, MD.

Ponder, D. (2019). "Frequency Wave Theory: A New Paradigm for Understanding the Universe." Presented at the International Conference on Theoretical Physics, Kyoto, Japan.

Greene, B. (2015). "The Future of String Theory." Keynote address at the Strings Conference, Princeton University.

Online Resources

Weinstein, E. (2020). "The Portal Podcast." Retrieved from https://www.ericweinstein.org/portal

Ponder, D. (2021). "Introduction to Frequency Wave Theory." Retrieved from https://www.frequencywavetheory.com

Stanford Encyclopedia of Philosophy. (2016). "Quantum Mechanics." Retrieved from https://plato.stanford.edu/entries/qm/

MIT OpenCourseWare. (2010). "General Relativity." Retrieved from https://ocw.mit.edu/courses/physics/8-962-general-relativity-spring-2006/

National Aeronautics and Space Administration (NASA). (2019). "Cosmology: The Study of the Universe." Retrieved from https://www.nasa.gov/mission_pages/hubble/story/cosmology.html

Journals

Physical Review Letters

Journal of High Energy Physics

Nature Physics

Reviews of Modern Physics

Proceedings of the National Academy of Sciences

Miscellaneous

Ponder, D. (2022). Personal communication on the development of Frequency Wave Theory.

Weinstein, E. (2018). "Interview on Geometric Unity." Conducted by the Science Weekly podcast.

This bibliography reflects a selection of the most influential and relevant works that have shaped the ideas and research presented in this book. It includes foundational texts, key papers, and contemporary resources that together provide a comprehensive background for understanding the integration of Geometric Unity and Frequency Wave Theory.

IMAGES

Thanks for reading!
#FrequencyWaveTheory
@drew_ponder

Made in the USA
Columbia, SC
01 September 2024

41374396R00041